The Cybersecurity Queen

Cyber Crown

Unveiling Women's Journey in the Queen's Path to Cybersecurity

Dedication

This book stands as a tribute to the remarkable women who have fearlessly ventured into the realm of cybersecurity and other traditionally male-dominated domains. Their bravery, relentless determination, and unyielding spirit in shattering barriers and following their dreams serve as a beacon of inspiration for us all. Their legacy is etched within this book, a legacy that fuels the fire of progress and equality.

This book also acknowledges the men who stand resolutely as allies and champions of diversity and inclusion. Their support and dedicated efforts in fostering environments of inclusivity are a cornerstone in our collective journey toward a more equitable future. They play a pivotal role in amplifying diverse voices, embracing distinct viewpoints, and nurturing collaborative spaces where every individual can flourish. Their commitment ignites positive change, shaping a world where all can shine.

Finally, this book is devoted to the young minds who are inquisitive and eager to venture into a world of diverse career prospects. May it kindle their creativity and infuse them with the courage to envision grand aspirations. As they absorb knowledge and flourish, they understand that the horizons of their achievements are boundless. Embrace their inherent potential and embark on a path illuminated by infinite opportunities. The trajectory of our shared destiny rests upon their brilliant intellects and empathetic spirits.

May this book be a beacon of unity, guiding us all towards a world where talent knows no boundaries, and where women and men work together in harmony to safeguard the digital crown. Let us, united as one, secure a future that celebrates diversity, inclusivity, and the boundless potential of every individual.

Acknowledgment

I would like to express my heartfelt gratitude to the following individuals and groups who have played instrumental roles in shaping my journey in the cybersecurity field.

First and foremost, I am deeply grateful to my village of powerful, committed and loving individuals. This includes my closest family and friends. Their commitment to fostering a supportive and encouraging environment has given me strength throughout my youth and adolescent years. Their inspiration, belief in my potential, and constant empowerment have been the driving force behind the pursuit of my dreams and goals. I am truly grateful for their positive influence and providing me with an example of what it means to work as a team. Thank you for being my foundation and for sharing this incredible journey with me.

I extend my sincere appreciation to the staff and faculty at the Pennsylvania State University, particularly the College of Information Sciences and Technology

(I.S.T.). The exceptional education and resources provided by the university have been invaluable in equipping me with the knowledge and skills necessary for a career in cybersecurity. I am indebted to my family and friends within the I.T. and cybersecurity field, who have created a vibrant community of knowledge-sharing and collaboration. Their insights, guidance, and camaraderie have been instrumental in my growth and development as a cybersecurity professional. I would also like to express my gratitude to all my managers and colleagues who have contributed to fostering an inclusive work environment. Their support, mentorship, and collaborative spirit have been instrumental in my personal and professional growth.

Lastly, I extend a special thanks to my Boss Mom, aka B.M.T., who has been an exceptional example, leader, and mentor. Her guidance, wisdom, and dedication to fostering a thriving cybersecurity community have been invaluable. To all those mentioned and countless others who have supported and inspired me on this journey, I am truly grateful. Your contributions have shaped my path and contributed to my success in the cybersecurity field. Thank you for being part of my journey and for being a source of inspiration and encouragement.

Table of Contents

Dedication

Acknowledgment

Preface

About the Cyber Crown

Who is the Cybersecurity Queen?

The World of Cybersecurity

    1.1 - Defining the Domain of Cybersecurity

    1.2 - The Spectrum of Cyber Threats

    1.3 - Common Vulnerabilities

    1.4 - The Impact of Cyberattacks

    1.5 - Developing Cybersecurity Skills

    1.6 - Cybersecurity Careers

    1.7 - Cybersecurity Best Practices

Women Trailblazers and Pioneers

    2.1 - Early Contributions and Trailblazers

    2.2 - Overcoming Gender Barriers

    2.3 - Inspiring Stories of Women in Cybersecurity

Women in Cybersecurity

    3.1 - Importance of Women in Cybersecurity

3.2 - Challenges Faced by Women in Cybersecurity

3.3 - Breaking Barriers and Shattering Stereotypes

3.4 - Inspiring Women Leaders in Cybersecurity

3.5 - Strategies for Advancement and Empowerment

3.6 - Building Supportive Communities

Diverse, Equity and Inclusive (D.E.I.) Culture

4.1 - Importance of D.E.I. in Cybersecurity

4.2 - Promoting an Inclusive Workplace

4.3 - Addressing Gender Bias and Discrimination

The Future of Women in Cybersecurity

5.1 - Emerging Trends and Technologies

5.2 - Opportunities and Challenges Ahead

5.3 - Nurturing the Next Generation

The Reign of Women in Cybersecurity

6.1 - Encouraging More Women to Join the Field

6.2 - Embracing the Power of U.N.I.T.Y.

Preface

Welcome to the dynamic realm of cybersecurity, where you will delve deep into the core of cybersecurity, comprehending its essence and gaining insights into the scope of threats and vulnerabilities that impact both individuals and organizations. This immersive journey will equip you with the knowledge of requisite skills and unveil an array of prospective careers while also laying out the bedrock of best practices for nurturing a robust security stance.

Embark on a voyage through the narratives of women, tracing their impactful journeys and contributions that have shaped the present landscape of the field. From pioneers who carved paths through history to pragmatic insights that resonate today, you'll embrace the spirit of diversity and inclusion, recognizing the profound unity between women and men in fortifying the digital castle of our world. Join us on this transformative journey, where resilience, empowerment, and guardianship intertwine, guiding us through the labyrinth of cybersecurity's ever-evolving terrain. Together, we carve a trajectory toward a future marked by enhanced safety, armored security, and a resolute unity of purpose.

Let the extraordinary journey commence.

About the Cyber Crown

"Cyber Crown: Unveiling Women's Journey in the Queen's Path of Cybersecurity" embarks on a comprehensive exploration of the cybersecurity realm, celebrating the diverse and empowering narratives of women. From its inception, this book accentuates the pivotal role that cybersecurity occupies in our rapidly digitizing world while also providing insights into fortifying digital defenses for both individuals and organizations. Within these pages, you'll gain a panoramic perspective of the evolving landscape and the critical importance of proactive readiness, ranging from fundamental knowledge and educational pursuits to the practical application of acquired skills.

The historical accounts of pioneering women and trailblazers serve as beacons of inspiration, honoring their past accomplishments and catalyzing future triumphs. These examples, just a glimpse of many, underscore the imperative for more women to carve their paths, making a resounding impact in the face of forthcoming challenges. Exploring the unique contributions of women within a predominantly male domain, you'll come to grasp the significance of diverse perspectives, skill sets, and determination that conquer a myriad of obstacles. The call for women is not limited to pursuing careers in cybersecurity but also to seize leadership roles, drive research, foster supportive communities, and elevate our digital defenses. By witnessing more exemplars of potential, the subsequent generation of cybersecurity experts will flourish in diversity and inclusivity, nurturing an environment where every idea and viewpoint is embraced.

As the narrative unfolds, the book delves into increasing trends and technological frontiers in cybersecurity, revealing avenues of opportunity alongside the challenges that lie ahead. It beckons women to embark upon their individual journeys within this domain, accentuating the indispensability of their voices and contributions. Moreover, it extends a call to men as allies, encouraging their active role in cultivating workplaces that are diverse and inclusive. The book's message resounds, urging readers to put on their metaphorical digital crowns and catalyze transformation within the dynamic landscape of cybersecurity.

Transcending gender and age boundaries, the Cybersecurity Queen herself requests each of you to embrace the immense power of knowledge, immersing yourselves wholeheartedly in the captivating realm of cybersecurity. A collective responsibility spanning expertise and privilege permeates every facet of our lives. In our interconnected world, where vulnerabilities reach all, recognizing the shared endeavor of safeguarding ourselves, our cherished ones, and our associates stands paramount. Armed with knowledge, attuned to advancements, and fortified with protective measures, an impervious shield emerges against the ceaselessly evolving cyber threats that loom on the horizon. Let the wisdom nestled within these pages be your steadfast defense, charting a course toward a future fortified by safety, security, and unshakeable digital fortitude. United in purpose, we shall surmount every obstacle, carving an unyielding path of resilience and safeguarding.

Who is the Cybersecurity Queen?

What if I shared with you that a single moment in the digital realm could have a profound impact on your reality? The journey of the Cybersecurity Queen, also known as Shari Mitchell, embarks on a captivating expedition that presents the crossroads between the cyber domain and tangible existence. This journey illuminates the paramount significance of knowledge, comprehension, and vigilance in shielding ourselves and our counterparts from the potential perils lurking within the digital expanse. As we delve into her narrative, we bear witness to the transformational potency of mentorship, personified by her remarkable guide and confidante, fondly known as "Boss Mom" (B.M.T.). Jointly, they navigate the intricate labyrinth of the cybersecurity landscape, unraveling insights, fostering evolution, and kindling inspiration for others to etch their imprint in this realm.

This journey began in 2005, an era marked by the ascension of social media, capturing her curiosity with the allure of digital connections. Within this domain, she confronts the shadowed facets of social media, a firsthand encounter with the aftermath of identity theft and the peril it begets. An unknown malicious actor used her pictures, personal information, and sensitive information about other students to create a fake profile. This profile caused a lot of disruption in her life and throughout the entire school. The Cybersecurity Queen anxiously pondered the identity of the perpetrator and the reasons behind their actions. After the school board investigated the incident, the perpetrator was expelled from school, and she was able to clear her reputation. Her resurgence from this identity breach fortifies her determination and amplifies her commitment to nurturing a more secure digital haven for all. This tale remains a sensitive testament to the resilience and tenacity that can sprout from adversity, nurturing personal growth and cultivating a safer and empowered online community.

Another lesson learned along the Cybersecurity Queen's journey exposes the unrelenting threats that lurk around the digital kingdom – malware attacks. As a passionate enthusiast of music, she ventures onto unvetted platforms that offer free downloads. Often, the downloads resulted in a safe and successful music file transfer. She hit the lottery when one file contained malware that encrypted all her local files. It was evident that she was battling a ransomware attack that had the potential to jeopardize her educational pursuits. It truly takes a village because a family friend was knowledgeable about technology and cybersecurity. He helped to recover her files while urging the importance of a healthy security posture that trusts only verified websites and includes antivirus and malware detection software. This encounter triggers a mission—to educate others about the hazards tethered to unverified platforms and the invaluable lessons of cultivating robust security practices. This episode propels her into the realm of cybersecurity, where she embraces the mantle of responsibility to safeguard herself and her counterparts from the veiled threats of the digital realm.

The Cybersecurity Queen deeply respects the role of education, which has been guided by an innate curiosity and an affinity for technology. She embarks upon a journey of perpetual learning and expansion by attaining a degree in Security and Risk Analysis from The Pennsylvania State University. Armed with knowledge and proficiency, she adeptly navigates the shifting terrain. Following her graduation, she embarks upon a tenure within the I.T. development program of a multibillion-dollar international technology powerhouse, acknowledged as the Best Place to Work in Austin in 2022. The fabric of her journey unfolds further as she ascends to new heights, reigning over a high-visibility customer engagement program that oversees the wide scope of cybersecurity. Here, she combines her technical prowess with a passion for cultivating relationships and bolstering the business. Through this conduit, trust flourishes between her company and its clients, elevating it to the pedestal of a respected advisor and an industry partner.

Mentorship threads the very fabric of the Cybersecurity Queen's success; thus, it is no revelation that she crosses paths with B.M.T., the quintessential archetype of an exemplary mentor. B.M.T.'s voyage from a corporate administrative assistant to a distinguished professional in cybersecurity serves as a compelling saga. Her journey encompassed challenges and successes that promoted compassion for her team, recognition of their capacities, and opportunities for growth. With a mentorship spirit that is nurturing, unwavering, and uplifting, she underscores the value of seeking wisdom from seasoned mentors, who, through their insights, experiences, and expertise, pave the path to safeguarding ourselves and our digital world. Her mentorship is a resounding proclamation of the advantage inherent in learning from those who have tread the course before us.

As the chronicle of the Cybersecurity Queen unfolds, let us ponder upon the significance of education, mentorship, hands on experience, and the essence of cybersecurity's role in our interconnected existence.

Together, we shall harness the might of knowledge, strengthen our cybersecurity measures, and remain ceaselessly vigilant, ensuring the sanctity of our digital identities and the welfare of ourselves, our loved ones, and our organizations.

Let us navigate the enigma of the cyber realm armed with caution, determination, and a commitment to fortifying our digital kingdoms.

The World of Cybersecurity

In the realm of the digital landscape, cybersecurity stands as a formidable fortress, encircling and safeguarding the cyber crown. It's a practice that involves implementing defenses to shield computer systems, networks, and data from cyber threats. Its purpose is to fend off unauthorized access, prevent theft and damage, and maintain the sanctity of your digital kingdom by upholding the pillars of confidentiality, integrity, and availability. In this intricate web of connectivity, individuals are inspired to pursue educational pathways that align with their passion and develop the skills to protect themselves through a combination of best practices that will promote a healthy security posture. It's imperative to understand how we interact with cybersecurity daily and the key things to look out for to best protect yourself and your loved ones from malicious actors.

1.1 - Defining the Domain of Cybersecurity

What does cybersecurity truly entail, and what significance does it hold? At its core, cybersecurity emerges as a guardian, a shield against unauthorized access, data theft, and digital harm. This mantle has become paramount in a world woven with intricate digital threads. Cybersecurity thrives on its multifaceted nature, standing guard against a spectrum of threats and vulnerabilities that seek to compromise the core values of digital assets - integrity, confidentiality, and availability. The weaponry used in this battle ranges from measures and protocols to thwart hacking, malware, phishing, and other cyberattacks. It's an active defense mechanism that ensures the sanctity of sensitive data, the operational longevity of systems, and the safety of critical infrastructures.

This defense strategy extends its reach across a wide variety of realms, be it government, finance, healthcare, education, and beyond. In fact, the embrace of technology and the storage of sensitive information render virtually every entity susceptible to the sway of cybersecurity's challenges. This unifying need underscores the importance of constructing robust strategies, harnessing advanced technologies, and cultivating a culture where cybersecurity is not just an afterthought but a living creed. As the digital frontier continues to evolve, the landscape morphing with every twist and turn, cybersecurity transforms into a relentless sentinel, adapting to emerging threats and constantly changing vectors of attack. Cybersecurity professionals deploy a plethora of tools, techniques, and practices to mitigate risks and shield digital assets. This relentless dance with danger mandates perpetual learning, staying abreast with the latest trends, and nurturing an anticipatory mindset, ensuring the advantage always remains with the defenders.

Understanding the scope and importance of the Cybersecurity domain empowers individuals and organizations to act with discernment with their steps guided by awareness and knowledge. The ramifications of cyber threats, such as financial losses, reputational blight, and privacy breaches, become vivid and real, propelling stakeholders to embolden cybersecurity within their strategies and decisions.

1.2 - The Spectrum of Cyber Threats

Malware, an array of malicious software, lurks as a pervasive specter that is designed to infiltrate, disrupt, and plunder systems and devices. This cyber threat includes viruses, worms, Trojans, and ransomware, each with its distinct capability. It's typically delivered through sinister emails, poisoned websites, or the guise of corrupted software. The countermeasure consists of a robust antivirus solution and regular system updates. The next type of cyber threat is the sly trickster we know as phishing. It's an act of deception where malicious actors pretend to be legitimate entities to gain trust. Like chameleons, they adopt various guises, be it email spoofing, fraudulent websites, or social engineering techniques. The guards against this deceit involve vigilance - to trust only what is known and verified. Understanding the telltale signs and mastering the fine art of phishing detection and mitigation becomes paramount. The dangers of social engineering, a type of phishing attack, lie where manipulation thrives and human vulnerabilities are exploited. Hackers use tactics such as pretexting, baiting, and tailgating techniques that target the human psyche. This psychological warfare gains access to confidential information that circumvents barriers through the art of trust. Recognizing the threads of manipulation woven by these malicious actors arms individuals and organizations with the wisdom to repel by cultivating a culture where every fiber is a sentinel.

Ransomware attacks are considered modern-day digital terrorist that encrypts data, demanding a large amount of money for its release. This type of cyberattack targets individuals, businesses, and even critical infrastructure. It can be distributed through malicious emails, exploited software, or corrupted websites. Vigilant precautions through data backups and robust security protocols stand guard against this attack, reducing the ransomware's potential impact. Another common cyber threat is Distributed Denial-of-Service (DDoS) attacks, where the floodgates of traffic are opened to drown the targeted victim. Techniques like botnets, amplification, and application layer strikes are used to disable the availability of data. The counterattack against this threat entails network monitoring, traffic filtration, and an architecture fortified to withstand the tempest, preserving the steady flow of digital services.

Within the realm of threats lurks the insidious insider, an individual who is cloaked in legitimacy while exploiting authorized access to orchestrate harm. Employees, contractors, business partners, and accomplices intentionally or inadvertently misuse their privileges to carry out malicious activity, including data breaches, theft, or sabotage. Proactive countermeasures to prevent insider threats involve conducting background checks, strict access controls, monitoring user activities, and fostering a security awareness and reporting culture. By addressing insider threats, organizations can mitigate risks and protect the integrity of sensitive information and assets from unauthorized access and misuse.

Supply chain attacks pose a significant threat to organizations as they target vulnerabilities within the interconnected network of suppliers and vendors. These attacks can occur at various stages of the software development lifecycle, including the design, development, distribution, and maintenance phases. Attackers exploit weak links in the supply chain to inject malicious code, tamper with software updates, or compromise the integrity of physical components. These attacks can lead to the distribution of compromised software, unauthorized access to systems, or the insertion of backdoors for future exploitation. To mitigate the risk of supply chain attacks, organizations must implement rigorous security measures, such as conducting thorough risk assessments, vetting suppliers and third-party components, and implementing strong security controls throughout the supply chain. By ensuring the integrity and security of the supply chain, organizations can safeguard their digital infrastructure and protect against the devastating consequences of supply chain attacks.

The multitude of cyber threats requires a layered defense to make it extremely difficult for malicious actors to compromise the integrity, availability, and confidentiality of data. By recognizing the tactics employed by these cybercriminals and embracing an informed stance, individuals and organizations forge an armor against cybercrime. It is a call for continuous learning, unified action, and proactive measures to combat the constantly evolving landscape of cyber threats.

1.3 - Common Vulnerabilities

Within the fabric of digital architecture, vulnerabilities are inevitable. They serve as weaknesses in the cyber armor that malicious actors can exploit through outdated systems, misconfigured networks, or human behavior. Cybercriminals exploit common vulnerabilities to disclose, alter, or destroy the data of an individual or company. Many of the well-known data breaches have happened due to a lack of proper security hygiene. It's best to understand each common vulnerability mentioned and the countermeasures to minimize the likelihood and impact of cyber threats.

Humans are, unfortunately, the weakest link to any system, which causes inherent vulnerabilities that make social engineering exploits extremely common. Cybercriminals often exploit human vulnerabilities, such as susceptibility to phishing attacks, sharing sensitive information, or falling victim to social engineering tactics. Organizations can empower their workforce to resist these manipulation techniques through security awareness, training, and education. Inadequate security practices could include poor security hygiene, such as weak or reused passwords, unauthorized software installations, and unsecured devices. Companies should have robust security policies and employee training programs to enforce security best practices throughout the organization. By promoting a culture of security and accountability, individuals and companies can strengthen their defense against cyber threats and minimize the potential impact of breaches.

Misconfigured systems and insecure network configurations result from common errors and oversights that can expose systems and devices to cyber threats. This includes weak passwords, default configurations, open ports, and misapplied security policies. Individuals and organizations should adopt best practices in system configuration and network security, such as implementing strong passwords, limiting access privileges, and conducting regular security audits.

There are also risks associated with insecure third-party components and supply chain vulnerabilities. Companies relying on third-party software or hardware may contain hidden vulnerabilities or backdoors that pose an inherent danger. Conducting due diligence when selecting and vetting third-party vendors is important, ensuring that security is a priority throughout the supply chain. Transparency and security assessments are encouraged to mitigate the risks associated with supply chain vulnerabilities. By identifying these weak points within digital infrastructure, software, and human behavior, organizations can fortify their systems and protect against common vulnerabilities. A holistic approach to cybersecurity encompasses technical measures, ongoing risk assessments, employee education, and proactive security practices. Practical strategies and tools for mitigating vulnerabilities include vulnerability assessments, penetration testing, and secure coding practices. The goal is to bolster the security of digital assets with a layered defense approach and contribute to the ongoing battle against cyber threats.

1.4 - The Impact of Cyberattacks

Cyberattacks have a disruptive effect on business and operational continuity. The cascading impact of successful cyber intrusions on critical infrastructure, supply chains, and daily business operations can result in system downtime, data loss, and service disruptions, affecting productivity, customer satisfaction, and overall business resilience. Cyberattacks also have disruptive effects on individuals. The psychological and emotional impact experienced by victims of cyberstalking, identity theft, and online harassment can impact the quality of life of an individual. They witness the invasion of privacy, loss of personal information, and distress caused by violating digital boundaries. There are broader societal implications of cyberattacks that can potentially impact national security, public safety, and critical infrastructure. Vulnerabilities can be exploited to disrupt essential services, compromise government systems, and undermine societal trust.

Several companies have faced significant reputational damage that impacts customer trust and brand image. For instance, Target, a retail giant, suffered a massive data breach in 2013, compromising around 40 million customer records and payment information. Malicious actors gained access to the company's network through a third-party vendor. Target customers lost trust, which has long-term impacts like financial loss from legal fees, settlements, regulatory fines, and improving security measures. Similarly, LastPass, a popular password management service, faced a breach in 2015 that exposed user email addresses and encrypted master passwords. Although the passwords were encrypted, it is only a matter of time before the encryption is broken. If passwords are discovered using this method, then malicious actors could infiltrate the user's account on another platform if they are using the same password. This resulted in LastPass users losing trust and confidence in the service's ability to safeguard their data. There were also financial losses due to the data breach recovery and the improvement of their security measures to protect against future threats.

Data breaches, ransom demands, and intellectual property theft result in staggering costs. Companies typically will issue reactive countermeasures like incident response, recovery efforts, regulatory fines, legal actions, and the loss of business opportunities. It is wise for organizations and individuals to implement proactive countermeasures to avoid financial losses that ultimately impact the economy. The consequences of cyberattacks are beyond the technical realm but can cause reputational damage like the erosion of trust and confidence, customer attrition, brand devaluation, and diminished market competitiveness. We are collectively responsible for securing the digital landscape by contributing to the resilience of our communities and nations.

1.5 - Developing Cybersecurity Skills

Many universities and colleges offer undergraduate programs specifically tailored to cybersecurity. These programs provide a comprehensive foundation in computer science, network security, cryptography, and other key areas. Students gain a deep understanding of the technical aspects of cybersecurity and the legal, ethical, and social implications of the field. These programs typically offer computer programming, operating systems, network administration, risk management, and incident response coursework. Some universities also offer specialized tracks within their computer science or information technology programs, focusing on cybersecurity. For individuals seeking advanced knowledge and specialized expertise in cybersecurity, graduate programs offer an excellent opportunity to expand their skills. Master's degree programs in cybersecurity delve into advanced topics such as advanced cryptography, digital forensics, secure software development, and network defense. These programs often include hands-on projects and research opportunities to enhance practical skills further. Additionally, some universities offer Ph.D. programs in cybersecurity for those interested in pursuing research and academic careers.

In addition to formal degree programs, numerous specialized courses and boot camps are available for individuals looking to gain targeted cybersecurity skills in a shorter time frame. These courses often focus on specific areas of cybersecurity, such as penetration testing, ethical hacking, secure coding, or cloud security. They provide intensive training, practical exercises, and real-world scenarios to help individuals quickly acquire in-demand skills. Many of these programs offer industry-recognized certifications upon completion, further enhancing the credentials of participants. Certifications play a significant role in cybersecurity, as they validate an individual's knowledge and skills in specific domains. There are several well-recognized certifications available that cover various aspects of cybersecurity. Some popular certifications include Certified Information Systems Security Professional (C.I.S.S.P.), Certified Ethical Hacker (C.E.H.), Certified Information Security Manager (C.I.S.M.), and CompTIA Security+. These certifications demonstrate proficiency in security management, ethical hacking, risk assessment, and foundational knowledge of cybersecurity concepts. Obtaining these certifications can enhance employability and professional advancement in the cybersecurity industry.

Given the fast-paced nature of the cybersecurity field, continuous learning and professional development are crucial. Professionals in the industry are encouraged to stay updated on the latest trends, threats, and technologies through conferences, workshops, webinars, and online courses. Many organizations and professional associations offer resources and training programs to support ongoing education and skill development. These opportunities allow individuals to stay ahead of emerging threats and acquire advanced knowledge to excel in their cybersecurity careers. Aspiring cybersecurity professionals should consider their goals, interests, and learning preferences when choosing an educational pathway. It is advisable to research and assess the curriculum, faculty expertise, industry partnerships, and opportunities for practical experience when selecting a program or certification.

Excelling in the dynamic and challenging field of cybersecurity requires a diverse skill set encompassing both technical competencies and non-technical skills. Most cybersecurity professionals are multidimensional and understand the importance of maintaining a healthy balance between technical and non-technical skills.

To secure digital infrastructures, individuals must obtain key technical competencies, such as understanding network protocols, architectures, and vulnerabilities. Proficiency in firewalls, intrusion detection systems, virtual private networks (VPNs), and secure network design enables cybersecurity professionals to protect networks against unauthorized access, data breaches, and other network-based threats. Knowledge of cryptographic algorithms, encryption techniques, and secure protocols is fundamental in safeguarding sensitive data. Cryptography ensures data confidentiality, integrity, and authenticity. It is applied in various aspects of cybersecurity, including secure communication, data protection, and authentication mechanisms. A rapid and effective response to cybersecurity incidents minimizes damage and mitigates potential threats. Cybersecurity professionals should be skilled in incident detection, analysis, containment, and recovery. Understanding incident response frameworks, forensic techniques, and incident handling protocols helps organizations efficiently manage and respond to security incidents. Ethical hacking, also known as penetration testing or vulnerability assessment, involves identifying and exploiting security vulnerabilities to assess the resilience of systems. Proficiency in ethical hacking tools and techniques enables cybersecurity professionals to identify system and device weaknesses before malicious actors can exploit them.

Cybersecurity professionals should also possess non-technical skills, such as strong critical thinking skills, to analyze complex problems, evaluate risks, and make sound decisions. They must assess potential threats, vulnerabilities, and consequences to develop effective strategies and countermeasures. The ability to identify, analyze, and solve problems is essential in cybersecurity. Cybersecurity professionals must be resourceful and adaptive in addressing emerging threats, vulnerabilities, and system breaches. They employ creative problem-solving techniques to develop innovative solutions and minimize the impact of cyber incidents. Effective communication skills are crucial for cybersecurity professionals to convey complex technical concepts to technical and non-technical stakeholders. Clear and concise communication facilitates collaboration, enhances incident response, and promotes cybersecurity awareness within organizations. Cybersecurity is a team effort that requires collaboration with colleagues, I.T. teams, management, and stakeholders. Strong teamwork skills enable professionals to work cohesively, share knowledge, leverage collective expertise, and respond effectively to cybersecurity challenges. Additionally, networking with professionals in the field and seeking mentorship can provide valuable guidance and insights for career growth.

The cybersecurity landscape evolves rapidly, requiring professionals to adapt to new technologies, emerging threats, and changing regulations. A commitment to continuous learning, staying updated on industry trends, and acquiring new skills is crucial for success in this field. The multidimensional nature of cybersecurity expertise can effectively protect individuals, organizations, and nations by combining technical expertise in network security, cryptography, incident response, and ethical hacking with critical thinking, problem-solving, communication, and teamwork skills. The growing demand for cybersecurity professionals and their critical role in protecting individuals, organizations, and nations from cyber threats make this an exciting and rewarding career path for those passionate about safeguarding the digital realm.

1.6 - Cybersecurity Careers

In the dynamic field of cybersecurity, there are numerous opportunities to excel in roles such as Security

Analyst, Penetration Tester, Cybersecurity Engineer, Security Consultant, Cybersecurity Manager, Digital Forensics Analyst, Cloud Security Specialist, and Incident Response Manager, to name a few. Security Analysts stand as vigilant guards who monitor and decipher the scope and impact of security incidents, response, and security tools. Key skills for this role include proficiency in identifying and analyzing potential threats, responding quickly and effectively, and familiarity with security tools like incident detection systems (I.D.S.), firewalls, and security information and event management (S.I.E.M.) solutions. Penetration Testers strengthen organizations' security posture by uncovering vulnerabilities through ethical hacking, vulnerability assessments to identify security weaknesses, and delivering concise vulnerability reports.

Cybersecurity Engineers are the architects who design secure solutions with expertise in secure software development, configuring secure networks, and utilizing encryption techniques. Security Consultants advise on security strategies with skills in risk assessment, compliance, navigating industry regulations, and strong communication to build lasting client relationships. Cybersecurity Managers lead teams with skills in inspiring diverse teams, managing security budgets, and developing long-term security strategies. Digital Forensics Analysts investigate cyber incidents, which require proficiency in specialized software and tools, adherence to legal protocols, and accurate evidence examination.

Cloud Security Specialists focus on securing cloud-based environments and specialize in understanding cloud architectures, managing identity and access controls, and staying up to date on the compliance of cloud security frameworks. Incident Response Managers lead teams that respond to cybersecurity incidents, which requires leading incident response efforts, staying composed under pressure, and documenting lessons learned from incidents for future improvements. In conclusion, the cybersecurity field offers a wide array of rewarding careers, each with unique challenges and opportunities. By acquiring and honing the key skills discussed in this chapter, those interested in pursuing a career in cybersecurity can forge successful paths and significantly contribute to the industry's advancement and protection of our digital world.

1.7 - Cybersecurity Best Practices

Below is a quick summary of the key best security practices that individuals of all ages should understand to better protect themselves against cyber threats and enhance their online safety and security. Malicious actors go through the path of least resistance, so arm yourself with the knowledge to improve your security hygiene to lessen the likelihood of becoming the next victim.

Always use strong and unique passwords! Regardless of age, everyone should create strong, unique passwords for their online accounts. A strong password should include upper and lowercase letters, numbers, and special characters (an example of a strong and unique password is VulN3r@B!li+ie$). It's best to use a different password for each account to reduce the risk of multiple accounts being compromised if one password is breached. Using weak passwords leaves one vulnerable to unauthorized access and potential compromise of personal and sensitive information. Always store passwords using a secure password manager! Sometimes, it's easier to write down passwords or put them in your phone's notepad. This poor security hygiene practice results in a higher risk of becoming a victim of a stolen identity or data breach. A better practice is to use secure password managers like NordPass. It can generate complex passwords, securely store passwords and passkeys through strengthened encryption, has a data breach scanner, and is user-friendly. Enable two-factor authentication whenever possible! This adds an extra layer of security by requiring a second form of verification, such as a temporary code sent to a mobile device, in addition to the password. The result of not enabling two-factor authentication is an increased risk of unauthorized access to accounts, as it relies solely on a single layer of security and makes it easier for malicious actors to compromise personal information.

Keep system and device software updated! These updates can seem annoying, but it is essential to regularly update operating systems, applications, and devices to ensure they have the latest security patches. Many updates include critical security fixes that protect against known vulnerabilities. The result of not keeping software up to date leaves vulnerabilities unpatched, which hackers can exploit to gain unauthorized access, compromise data, or carry out malicious activities on the system or device. Always foster safe browsing habits! Safe browsing practices include avoiding suspicious websites and clicking only links from trusted sources. Users should always look for the padlock symbol and "https://" in the website URL, indicating a secure connection. Unsafe browsing habits can expose individuals to malicious websites, online scams, and potential malware infections.

Develop healthy habits when receiving emails! Let's explore an example of a fraudulent sender like janedoe@bofa.gui.com. For starters, examine the sender's email address, company logo, links, and attachments carefully. Look for misspellings, grammar errors, unusual domain names, or inconsistencies. Inspect the domain by hovering your cursor over any links in the email without clicking. Check the domain and logo matches the official website. Fraudulent emails often use deceptive links to direct you to phishing sites. Genuine emails usually address the recipient by name, so be cautious of any generic greetings like "Dear Customer." Any language within the body of the email that signifies urgency, threatening language, or grammar errors is an immediate red flag. Do not send any personal information through email. Personal information can include, but is not limited to, SSN, address, phone number, or payment information. Always keep your operating system updated to protect against known vulnerabilities. When in doubt, trust your instincts! If something feels off about an email, trust your intuition and verify its authenticity through official channels.

Always update your privacy settings! The default configurations usually do not include a healthy security posture. Updating your privacy settings on all social media platforms and other online services will limit the amount of personal information shared publicly and can manage access to who can view the posts. Not updating privacy settings can lead to unauthorized access to personal information, privacy breaches, and increased vulnerability to online threats. Frequently back up important data! Regularly back up important files and data to an external storage device or a secure cloud service. This protects against data loss due to ransomware attacks, hardware failures, or other unforeseen events.

Be cautious of social engineering attempts! Individuals, especially seniors, need to understand that social engineering techniques are used to trick people into revealing sensitive information. Always be cautious when sharing personal information over the phone or in response to unsolicited requests. Only use secure Wi-Fi networks! Always use strong, unique passwords for Wi-Fi networks to prevent unauthorized access. Exercise extreme caution when using public Wi-Fi networks for sensitive activities like online banking or accessing personal accounts. Regularly monitor all your accounts! Everyone should regularly monitor their financial accounts, credit reports, and online activities for suspicious or unauthorized transactions. Prompt reporting of any unusual activity can help mitigate potential damage.

Leveraging best security practices is crucial for improving your security posture by safeguarding personal information, protecting against cyber threats, and maintaining a secure digital presence to ensure a safer and more resilient online experience for individuals of all ages. Remember, malicious actors like easy targets, so implementing all these best practices encourages a layered defense around your digital kingdom.

Women Trailblazers and Pioneers

Women have always been at the forefront as early contributors and trailblazers in cybersecurity. They have achieved remarkable advancements while laying the groundwork for the field, often against societal norms and prevailing gender biases. These pioneers defied expectations, entering a predominantly male domain and making groundbreaking strides in computer science, cryptography, cyber defense, privacy, and much more! Their groundbreaking work forms the bedrock of the modern cybersecurity landscape.

2.1 - Early Contributions and Trailblazers

One such trailblazer was Ada Lovelace, often considered the world's first computer programmer. In the mid-19th century, Lovelace worked with Charles Babbage on his Analytical Engine, a precursor to the modern computer. Lovelace's visionary insights into the machine's potential led her to write the first algorithm, making her a pioneer in computer programming. Another pioneer by the name of Grace Hopper was a computer scientist and United States Navy rear admiral. Hopper played a pivotal role in developing early programming languages, including COBOL. Her contributions to computer programming and relentless pursuit of innovation laid the foundation for modern software development.

Dorothy Vaughan, Mary Jackson, and Katherine Johnson were three African-American women mathematicians who worked at NASA during the space race. Their flawless mathematical calculations and programming skills were integral to the success of numerous space missions. Their story, depicted in the book "Hidden Figures" and its subsequent film adaptation, showcases these women's immense talent and determination to overcome racial and gender barriers. Moving into cryptography, Joan Clarke was a codebreaker during World War II. Clarke worked alongside Alan Turing and other brilliant minds at Bletchley Park, where they deciphered encrypted messages produced by the German Enigma machine. Clarke's expertise in solving complex codes and exceptional intellect made her an invaluable asset in the fight against the Axis powers.

Meet Elizabeth "Liz" Adams, a pioneering figure in the cybersecurity realm who made groundbreaking contributions to computer security, particularly in penetration testing, which set her apart as a true trailblazer. Her innovative techniques and determination challenged the norms, reshaping the cybersecurity landscape and laying the foundation for modern security practices. Her legacy serves as a testament to the transformative power of women's leadership in safeguarding our digital world and forging a path toward a more secure and inclusive future. Next up is Barbara Liskov, who was a computer scientist and recipient of the Turing Award. Liskov's research on programming languages and software engineering principles developed important concepts like the Liskov Substitution Principle. Her work laid the foundation for object-oriented programming (OOP) and has influenced countless developers and engineers in cybersecurity. OOP is a computer programming model that organizes software design around data and objects rather than functions and logic. Another trailblazer is Anita Borg, a computer scientist and founder of the Institute for Women and Technology. Borg's advocacy for gender diversity in the tech industry and her efforts to create spaces for women to thrive have profoundly impacted cybersecurity.

These are just a few of many trailblazers who shattered glass ceilings, defied societal expectations, and paved the way for future generations of women in cybersecurity. Their courage, resilience, and innovation continue to inspire and empower women in the field, reminding us of the importance of diversity and inclusion in creating a secure and equitable digital future.

2.2 - Overcoming Gender Barriers

Women regularly encounter biases, stereotypes, and prejudices, both overt and subtle, on their path to success. Women have displayed remarkable resilience and determination as they persevered, challenging societal norms and discovering their place in a male-dominated industry. To unravel the gender biases prevalent in society that created obstacles for women entering the field of cybersecurity, we must address gender barriers. Stereotypes and preconceived notions about women's abilities in technical fields cast doubt on their competence and undermine their potential. Despite these challenges, women pioneer in cybersecurity embraced their passion for technology and refused to be deterred by societal expectations.

Mary Ann Davidson was a prominent cybersecurity executive and advocate for women in the field. Davidson's experience with navigating a male-dominated industry provides valuable insights into the biases women face. Her perseverance in the face of adversity and commitment to excellence paved the way for future generations of women to thrive in cybersecurity. Susan Landau is a renowned cryptographer and privacy expert. Landau's journey exemplifies the tenacity required to challenge the status quo. She encountered resistance and skepticism in her pursuit of a career in cybersecurity. Still, she remained steadfast in her commitment to her craft. Landau's expertise and dedication eventually earned her recognition and respect within the industry.

Systemic biases and barriers have hindered women's progress in cybersecurity. Creating inclusive work environments that value diverse perspectives and contributions is imperative. Efforts by organizations and advocacy groups to promote gender diversity and equal opportunities in the field are highlighted, showcasing the strides made to address the gender gap. This is a testament to the resilience and determination of women who refused to be defined by societal expectations. Their stories of triumph inspire and motivate aspiring women in cybersecurity. From defying stereotypes to challenging systemic barriers, these women embody the spirit of resilience and demonstrate that gender should never limit one's potential. There is an emphasis on the importance of continued efforts to dismantle gender biases and promote inclusivity in cybersecurity. By recognizing the accomplishments and struggles of women pioneers, society can work towards a more equitable and diverse future.

2.3 - Inspiring Stories of Women in Cybersecurity

Despite the many challenges, women have excelled in various roles within the cybersecurity field. Their tenacious spirit showcases their accomplishments, innovations, and impact in protecting individuals, organizations, and nations from cyber threats.

Cynthia Dwork, an influential researcher and affiliate professor at Harvard University, has made significant contributions to computer science, cryptography, and cybersecurity. Her work in privacy-preserving data analysis and non-malleable cryptography has reshaped the security landscape. With enduring ambition and passion, she inspires women in technology to dream big and overcome challenges to make a lasting impact on the world. Chanathip Namprempre, a distinguished faculty member at Thammasat University in Thailand, is an expert in cryptography, computer, and network security, and distributed systems. Her passion for empowering her students, especially those from less privileged backgrounds, drives her to create opportunities and impart valuable knowledge. Chanathip encourages real dialogues to promote progress for women in the male-dominated field. Her advice to aspiring girls and young women is to keep an open mind, seek supportive environments, and prioritize working with good people to excel in cryptography and related careers.

Window Snyder, a cybersecurity trailblazer, has made a profound impact on the digital world. From her early days at @stake to later working with Microsoft and Apple, she revolutionized security by integrating it into software development. Her methodologies, like threat modeling and the security development lifecycle, pioneered cybersecurity practices now standard in the industry. Snyder's resilience and determination have set a precedent for women in cybersecurity. Her legacy serves as an inspiration to overcome challenges, push boundaries, and make a lasting impact on the cybersecurity landscape. Chani Simms, an award-winning cybersecurity leader and founder of SHe CISO Exec., has left a significant mark on the cybersecurity industry. As a licensed computer hacking forensics investigator, she works with subject matter experts (SMEs) as a virtual CISO, data protection officer, and Cyber Essentials Assessor. Through her platform, SHe CISO Exec., she empowers information security professionals to become emotionally intelligent cybersecurity leaders, making a profound impact on the industry's growth and development.

Countless stories exemplify the immense talent, dedication, and innovation present within the female cybersecurity community. These women have shattered glass ceilings, challenged stereotypes, and left an indelible mark on the industry. There are many women whose stories delve into emerging technologies, policy advocacy, and the evolving challenges of securing the digital world. Through these narratives, women have the transformative power in cybersecurity and continue to be inspired to forge their path in this dynamic and vital field.

Women in Cybersecurity

Throughout the Cybersecurity Queen's journey, she was privileged to collaborate with teams that championed inclusivity regardless of one's background. She personally experienced the profound impact of an inclusive company culture, intentional networking, and powerful mentorship that evolves into sponsorship. While her journey faced unique challenges, the presence of a supportive community has paved the way for her to be recognized for equal opportunities and career growth.

Women emerge as indispensable architects who shape the foundation of solutions that safeguard our digital world. Women bring diverse perspectives, creative ingenuity, and adept problem-solving skills that fortify cyber defense against evolving threats. The inspirational journeys of pioneers and trailblazers serve as a stepping stone in the cybersecurity pipeline and provide a blueprint for resilience as we continue to shatter barriers. These narratives underscore the transformative power of representation, mentorship, and supportive networks in cultivating an inclusive work environment that empowers women to excel. While strides have been made toward equal opportunities, a forward-looking shift is needed to encourage more women to thrive in male-dominated fields. Women's expertise and influence must be nurtured at each stage of the pipeline, from hiring to leadership roles. Retaining valuable talent relies on fostering a company culture that champions diversity, inclusivity, and the dismantling of barriers. Flourishing women leaders become guiding lights that illuminate the path to professional growth, talent retention, and supportive communities.

3.1 - Importance of Women in Cybersecurity

Women are significant in shaping the trajectory of cybersecurity technology as they drive positive change that will enable the resilience of an organization. As technology becomes increasingly integrated into every facet of modern life, the active participation of women is vital to harness its full potential and ensure that diverse perspectives influence innovation. In the realm of cybersecurity, women's contributions are invaluable. Women bring a multifaceted approach through their unique insights and problem-solving skills that can be life-changing when protecting the digital kingdom against the evolving nature of cyber threats. Their involvement improves the collective ability to anticipate, detect, and counteract evolving cyber dangers, safeguarding individuals, organizations, and societies at large.

By breaking traditional gender barriers, women in technology and cybersecurity challenge stereotypes and inspire future generations. Their accomplishments illuminate pathways for other women and girls while fostering an environment of inclusivity and opportunity. The diverse perspective of women sparks innovation that leads to more comprehensive and effective solutions. Women leaders in the industry translate into tangible benefits like the business performing better and exhibiting heightened resilience. Women's presence at the decision-making table enhances problem-solving, encourages creative thinking, and fosters a workplace culture that values different perspectives.

In essence, the strides made by women in the field resonate far beyond individual achievements. They amplify the industry's potential for innovation, enhance cyber defenses, and contribute to a more equitable and dynamic technological landscape. By embracing and supporting women's roles in these fields, we not only pave the way for a more secure digital future but also tap into a wellspring of talent and creativity that enriches our collective progress.

3.2 - Challenges Faced by Women in Cybersecurity

Women pursuing careers in male-dominated fields such as cybersecurity encounter a unique set of challenges that stem from deeply entrenched gender biases and disparities. In this complex landscape, gender-based obstacles intersect with professional barriers, creating a multifaceted set of difficulties. One of the primary challenges is pervasive gender stereotypes that cast doubts on women's technical aptitude and suitability for roles within the field. These stereotypes can lead to a hostile environment, where women often struggle with skepticism, a lack of support, and dismissive attitudes from male colleagues and superiors.

The underrepresentation of women in cybersecurity exacerbates these challenges. The scarcity of visible female role models and mentors restricts aspiring professionals' ability to envision a path forward. This lack of representation can undermine women's aspirations and self-confidence, making it harder to visualize themselves succeeding in these careers. Unequal opportunities in terms of career advancement, leadership roles, and project assignments persistently plague women in cybersecurity. Biases in hiring and promotion processes perpetuate gender disparities, denying women the chance to showcase their abilities and reach their full potential. This inequity is often reinforced by exclusionary dynamics within established industry networks, hampering women's efforts to build robust professional connections.

Balancing demanding work schedules with personal commitments is another significant hurdle. The relentless nature of cybersecurity, with its long hours and high-pressure environment, can lead to burnout, especially when coupled with family or personal responsibilities. These challenges are amplified for women with intersecting identities, such as those from diverse racial, ethnic, socioeconomic, or LGBTQ+ backgrounds. Their experiences are shaped by an intricate interplay of biases and systemic barriers that demand a comprehensive and inclusive approach to dismantling inequality. While the road may be arduous, understanding and addressing these challenges is crucial to cultivating a more inclusive and equitable environment within male-dominated fields like cybersecurity. Acknowledging the unique challenges women face is an essential step towards breaking down these barriers and empowering women to thrive in these critical industries.

3.3 - Breaking Barriers and Shattering Stereotypes

Breaking barriers and shattering stereotypes in the field of cybersecurity holds profound significance in fostering a more inclusive, innovative, and resilient industry. As a traditionally male-dominated domain, cybersecurity has long been marred by gender biases and preconceived notions about women's capabilities. By challenging and overcoming these entrenched stereotypes, women are not only proving their competence but also reshaping the narrative of what a cybersecurity professional can be. These trailblazers inspire a new generation of aspiring women to pursue careers in cybersecurity, infusing the field with diverse perspectives and fresh ideas.

The act of shattering stereotypes not only empowers individual women to reach their full potential but also contributes to a more dynamic and effective industry. As women ascend to leadership positions, engage in pioneering research, and drive technological innovation, they showcase the power of determination, skill, and creativity in confronting challenges. Their success not only paves the way for others but also contributes to dismantling systemic biases and fostering an environment where everyone's contributions are recognized and valued. Ultimately, by breaking barriers and shattering stereotypes, women are propelling the cybersecurity field toward a future defined by excellence, equality, and limitless possibilities.

3.4 - Inspiring Women Leaders in Cybersecurity

Effective leadership in a male-dominated field like cybersecurity requires a unique blend of skills and attributes that can empower women to not only excel but also inspire others to follow suit. An effective leader must possess strong communication skills, the ability to collaborate effectively, and a strategic mindset to navigate complex challenges. These skills can be leveraged by women to gain an advantage in the field, enabling them to forge connections, break down barriers, and drive innovation. By fostering open communication and building relationships, women leaders can bridge gender gaps and create a more inclusive work environment that encourages diverse perspectives.

Strategic thinking allows women leaders to anticipate industry trends and position themselves as trailblazers, gaining respect and influence. Furthermore, cultivating resilience is essential to persevere through obstacles and setbacks, which is crucial in a field that demands continuous adaptation. Women who embrace these skills can rise as role models and inspire fellow aspiring women to take the leap into leadership roles. By showcasing their achievements and sharing their journeys, they provide tangible evidence that success is attainable, instilling confidence and motivation in others.

Women leaders in cybersecurity who break through barriers become beacons of hope for those who follow, demonstrating that gender should never limit one's aspirations. These leaders not only contribute to the advancement of the field but also pave the way for a more inclusive and diverse future. By nurturing the next generation, women leaders create a cycle of empowerment, where each success story fuels the aspirations of another aspiring woman. Ultimately, the journey from professional to leader in cybersecurity becomes a powerful narrative of overcoming challenges, dismantling biases, and achieving personal and professional fulfillment.

3.5 - Strategies for Advancement and Empowerment

In the rapidly evolving realm of cybersecurity, achieving career advancement and empowerment requires a strategic approach and a proactive mindset. To excel in this dynamic field, it is essential to continuously update your skills and knowledge through certifications, workshops, and conferences. Developing leadership abilities is key—take on leadership roles (official or unofficial), hone communication and decision-making skills, and seek mentorship to guide your growth.

Networking is a cornerstone of success. Build a strong professional network by engaging with industry associations, attending events, and participating in online communities. Seek mentors and role models who can provide guidance and inspiration as you navigate your career path. Challenge stereotypes and biases by advocating for diversity and inclusivity, fostering a supportive environment for everyone. Set clear career goals and step out of your comfort zone to embrace challenges and seize new opportunities. Effective communication, both within the technical realm and for non-technical stakeholders, is vital. Support others by becoming a mentor and contributing to a sustainably collaborative cybersecurity community. Prioritize self-care and maintain a healthy work-life balance to sustain long-term success.

Advocate for inclusive policies in your workplace and industry to promote and encourage equal opportunities for all. Embrace setbacks as opportunities for growth and persevere with resilience. By following these practical strategies, you can navigate the complex cybersecurity landscape, break barriers, and contribute to a diverse, inclusive, and empowered industry. Your journey may have challenges, but with determination and the right tools, you can achieve meaningful career advancement and make a lasting impact.

3.6 - Building Supportive Communities

Women have a plethora of communities and networks available to connect with like-minded professionals. These groups offer empowering platforms dedicated to the recruitment, retention, and advancement of women in cybersecurity. Women in Cybersecurity (WiCyS) brings together aspiring and thriving women professionals globally. This organization provides opportunities for collaboration, knowledge sharing, networking, and mentorship through conferences, career fairs, and development programs. The Women's Society of Cyberjutsu (WSC) focuses on closing the gender gap in information security roles by offering empowerment programs, hands-on training, education, and mentoring to help women succeed in the industry. CybHER's priority is to increase diversity within the field by encouraging girls (starting at middle school) and women professionals to explore cybersecurity through resources and support. Additionally, Women of Cybersecurity (WoSEC) emphasizes supporting one another and shining a light on accomplishments within the community. Other networks, such as Women in Security and Privacy (WISP), advance women and underrepresented communities in security and privacy through education, mentoring, and leadership opportunities.

The SANS Women's Immersion Academy provides scholarship-based, intensive cybersecurity training and certifications that enable successful career launches. Code Like a Girl celebrates women in technology by amplifying their voices and encouraging participation through technical articles and stories. Cyversity addresses the lack of diversity in cybersecurity through scholarship opportunities, mentoring, and workforce development programs. Women Leading Privacy fosters support and advancement for women in the privacy profession. InfoSecGirls encourages women's participation in information security events. Lastly, the Women Tech Network cultivates a global community of women in tech through leadership development, professional growth, mentorship programs, and networking events. These diverse communities play a vital role in empowering women to succeed and thrive in the dynamic world of cybersecurity.

Diverse, Equity and Inclusive (D.E.I.) Culture

Diversity is the invitation to the party, inclusion is the offer to dance, and equity is the appreciation of everyone's contribution, enhancing the overall harmony of the group. The commitment to diversity and inclusion in cybersecurity is recognized through ongoing strategies and initiatives. Organizations, driven by the recognition of its importance, are orchestrating programs, policies, and practices aimed at cultivating a more diverse and inclusive community. The recruitment and hiring landscape is undergoing transformation, shedding biases and welcoming a spectrum of talent. But the journey doesn't stop there. The stage is set for mentoring to flourish, nurturing the growth of underrepresented groups and providing them with the skills and knowledge to take center stage. Leadership development programs are akin to dance classes, refining the moves of future leaders and preparing them to lead the ensemble with confidence.

However, the ball is still in play, and the battle against gender bias and discrimination persists. The cybersecurity arena must remain vigilant by ensuring all qualified individuals have equal access to opportunities and are evaluated solely on their merits and performance. Just as a dance routine is perfected through practice, the cybersecurity industry must continuously strive to improve its diversity and inclusion practices. The more harmonious the dance, the stronger the collective defense against cyber threats. Ultimately, by nurturing diversity, fostering inclusion, and upholding equity, the organization will not only defeat digital adversaries but also create a richer, more innovative, and resilient future.

4.1 - Importance of D.E.I. in Cybersecurity

Employees who feel valued and respected for their unique contributions are more likely to be motivated, productive, and committed to their work. Diversity in male-dominated industries ensures that a wide range of perspectives are considered. By bringing together individuals from different cultural, ethnic, and social backgrounds, diverse teams can approach problems from various angles, fostering innovative thinking and creative solutions. Different perspectives challenge conventional wisdom and offer fresh insights, leading to more robust cybersecurity strategies. When teams encompass individuals with diverse experiences and skill sets, they can draw upon a wider pool of ideas and approaches.

Diverse teams can identify novel strategies, unconventional patterns, and alternative solutions that may have been overlooked in homogeneous environments. Cybersecurity challenges are multifaceted and continue to evolve, which requires comprehensive approaches. Diverse teams contribute to a more in-depth understanding of threats by leveraging their varied experiences and expertise. They can consider a broader range of attack vectors, anticipate emerging trends, and develop holistic solutions that address the unique needs of different user groups.

Diversity fosters empathy and promotes user-centric design in the development of cybersecurity methodologies. By incorporating diverse perspectives, teams can better understand diverse user groups' needs, behaviors, and vulnerabilities. This enables the development of cybersecurity solutions that are inclusive, accessible, and effective for all individuals, regardless of their backgrounds. The user-centric design ensures cybersecurity measures consider the human element, making them more secure, impactful, and user-friendly. Cybersecurity is a global issue that transcends borders. Diversity within the field reflects the diversity of the global community it seeks to protect. By embracing diversity, the field can better understand the unique challenges different regions, industries, and user groups face. It enables the development of solutions that are culturally sensitive, responsive to local contexts, and globally relevant.

Embracing diversity is not just a matter of inclusivity; it is essential for addressing the complexity and magnitude of cyber threats. The field can tap into a wealth of knowledge, perspectives, and experiences by fostering an inclusive environment that values and embraces diversity. This drives innovation, strengthens problem-solving capabilities, and ultimately enhances the cybersecurity posture across organizations and societies.

4.2 - Promoting an Inclusive Workplace

Male allies play a pivotal role in promoting inclusivity and cultural progress within the workplace. By recognizing the significance of diverse perspectives, men can contribute to a richer, more innovative environment. Recruiting and hiring are the foundation for supporting diverse candidates and ensuring a broad spectrum of perspectives join the workforce. As leaders, men should champion development programs that nurture underrepresented talent by providing them with the tools and guidance to flourish.

The support of male counterparts in leadership programs, sponsorship, and mentorship are transformative tools. These initiatives advocate for equal access and opportunity that empowers individuals from all backgrounds to ascend to leadership roles. Sponsorship is an essential bridge that influential men can use to advocate for deserving individuals, amplifying their achievements and opening doors for their advancement. Women need men to foster inclusive workplace cultures, which requires daily effort to create safe spaces for open dialogue where everyone's ideas are valued. Men who encourage cross-team collaborations allow for diverse skill sets to harmonize. Men can offer their guidance, share experiences, and foster an environment of mutual learning through mentorship.

The most important action for men as allies is to continue to challenge bias and discriminatory behavior, resulting in workplaces that are free from hostility. By actively participating in diversity and inclusion initiatives, men set an example for others to follow. Being committed to inclusivity is an investment in a stronger, more united workforce where every individual's potential is recognized and celebrated. Together, we forge a path toward a brighter, more inclusive future.

4.3 - Addressing Gender Bias and Discrimination

To the young and curious minds considering career paths in fields like cybersecurity, you are urged to recognize the potential within you that knows no bounds. Identify your passion and strengths that unleash your determination as your resilient armor. Equip yourself with the skills needed for a career field that excites you. Align yourself with a school or work environment that champions diversity and embraces individual uniqueness.

There has been great progress in ensuring inclusivity, but it's crucial to acknowledge the persistence of biases and discrimination. Gender bias, deeply rooted in preconceived notions, may seek to define your capabilities based on identity. However, the message stands strong – you are free to shape your aspirations and abilities. As your journey unfolds, you may encounter those who underestimate your capabilities. Lean on your strengths, skills, and support network to break through barriers and shatter stereotypes.

Success is measured when preparation meets an opportunity. It is crucial to remember that preparation is gauged by your time spent on developing skills, dedication to projects aligned with your passion, and creativity to spark innovation. As inclusivity evolves, more opportunities become available for future professionals to chase their dreams and gain tangible experience. Your contributions are valuable, irrespective of differences. Take challenges as opportunities to learn lessons, develop skills, and make immeasurable impacts in the digital kingdom.

The Future of Women in Cybersecurity

Women play a crucial role in shaping the foundation of solutions for the complex challenges posed by emerging technologies. Industries are currently evolving their internal processes to be compatible with emerging trends and technologies like artificial intelligence, machine learning, blockchain, the Internet of Things (IoT), the metaverse, and cloud security. Organizations will experience challenges that highlight the need to improve operational efficiency, enhance threat detection, and implement proactive defense mechanisms. The continuation of building diverse communities and inclusive workforce can attract and retain more female talent and leadership as technology advances.

5.1 - Emerging Trends and Technologies

Artificial Intelligence (AI) refers to the simulation of human intelligence that is typically processed by machines, particularly computer systems. These systems are designed to perform tasks that typically require human intelligence, such as understanding natural language, recognizing patterns, solving problems, and learning from experiences. AI involves various techniques like machine learning, where algorithms enable computers to learn from data and improve their performance over time. This emerging technology has diverse applications across industries, from virtual assistants and autonomous vehicles to healthcare diagnostics and financial forecasting, making it a transformative force in modern technology. AI is revolutionizing Cybersecurity by enabling advanced threat detection, predictive analytics, and automated response capabilities. More women must be at the forefront of AI-driven cybersecurity solutions, from developing innovative algorithms and models to identifying patterns for anomaly detection and building a proactive defense against evolving threats. Their diverse expertise and contributions are critical in harnessing the power of AI for effective cybersecurity defense. This emerging technology also has the potential to contribute positively towards addressing some key challenges women face, like gender bias and discrimination. The technology is neutral, but its application and the data used to train AI models can introduce biases if not handled carefully. AI can be leveraged to identify and mitigate existing biases in various domains, such as recruitment, criminal justice, or lending.

Machine learning techniques enhance emerging technologies by enabling systems to continuously learn, adapt, and detect sophisticated threats. A small percentage of women are actively involved in developing machine-learning algorithms that improve anomaly detection, behavior analysis, and malware identification. Their contributions are instrumental in building robust and adaptive defenses to stay ahead of cyber threats in today's dynamic digital landscape. More women are needed to be actively involved in exploring the applications of blockchain technology that contribute to Cybersecurity in various domains. With expertise in cryptography, decentralized systems, consensus algorithms, and interdisciplinary skills, women can prove to be valuable assets in developing secure solutions. In the context of IoT, women's contributions to developing secure architectures, protocols, and authentication mechanisms help address the unique cybersecurity challenges presented by the proliferation of connected devices. This work is essential for ensuring IoT systems' privacy, integrity, and availability. Similarly, in cloud computing, women's contributions to encryption mechanisms, access control frameworks, and security monitoring tools play a vital role in safeguarding sensitive data and mitigating risks associated with cloud-based cyber threats.

The Fourth Industrial Revolution (4IR) signifies the ongoing fusion of digital, biological, and physical innovations that will reshape industries and societies like the metaverse concept. This era embraces a combination of technologies like artificial intelligence, robotics, the Internet of Things, and blockchain that foster unprecedented connectivity and automation. The metaverse emerges as a digital space of exploration and interaction that combines virtual and augmented reality elements. It offers users a platform to "live" within a digital universe that allows them to engage with virtual environments through technologies like virtual reality (V.R.) and augmented reality (A.R.). This realm holds great potential for collaboration, creativity, and immersive experiences, with real-world applications spanning education, gaming, fitness, and more. As the 4IR advances, the metaverse's development and inclusion remain pivotal in shaping the future of technology and human interaction.

Despite being active and engaged users of the metaverse, women remain systematically excluded from influential positions that support the technology. While women have shown dedication and initiative in the metaverse, they still struggle to secure executive positions, thereby emphasizing the need to address these systemic inequalities. The metaverse's potential lies in its limitless applications, yet women's involvement in shaping its direction is disproportionately low. As this digital realm evolves, it is crucial to break the barriers that obstruct women's entry into leadership roles. For lasting change, the tech industry must actively recruit women for participation and leadership in new platforms and technologies. Early involvement can eliminate gender biases ingrained in project development. Women have a role to play by demanding participation in metaverse-building endeavors. Initiatives like involving more women and girls in the early stages of AI design and development can bring innovation and creativity to products that cater to a diverse audience. Bridging the gender gap in the metaverse is achievable through collective efforts, including diverse leadership, comprehensive audit of employment practices, visibility, and active recruitment of women. As the metaverse continues to unfold, embracing women's leadership and perspectives is essential for its success and evolution.

The transformative potential of emerging trends and technologies in Cybersecurity is vast, and women play a pivotal role in shaping and advancing these fields. Their expertise, innovative thinking, and interdisciplinary knowledge contribute to developing effective solutions that address the evolving cyber threat landscape. Women drive advancements that strengthen the overall cybersecurity posture by actively participating in research, development, and implementation efforts. To fully leverage the benefits of these emerging technologies, promoting diversity and inclusion in Cybersecurity is crucial. Encouraging women's participation and providing equal opportunities for their engagement and leadership fosters an environment where diverse perspectives, expertise, and creativity thrive. By embracing the contributions of women, the cybersecurity community can unlock the full potential of emerging trends and technologies, leading to more secure and resilient digital ecosystems.

5.2 - Opportunities and Challenges Ahead

The rapid advancement of emerging trends and technologies in Cybersecurity presents both opportunities and challenges for individuals and organizations seeking to protect digital systems and combat cyber threats.

- ➢ **Key opportunities in the field of Cybersecurity include:**

    Emerging trends and technologies offer the potential to enhance threat detection and response by analyzing vast amounts of data to identify patterns and anomalies, enabling quicker identification of cyber threats. These technologies promote better security of data through decentralized and tamper-resistance, making it suitable for identity verification and secure transactions. The robust protection of interconnected devices presents many challenges, but there are opportunities for expertise in safeguarding against potential breaches. Immersive digital environments call for novel security solutions to protect users' personal information and interactions. Lastly, more data and applications are migrating to cloud platforms, which require advanced protection mechanisms to defend against unauthorized access and data breaches. These emerging technologies present exciting prospects for bolstering cybersecurity and fortifying digital landscapes against evolving threats.

- ➢ **Key challenges in the field of Cybersecurity include:**

The risk of malicious actors exploiting these emerging trends and technologies is harder to detect as they leverage sophisticated measures like automation to enhance their attacks. Some emerging technology is designed to enhance data security but can also introduce challenges with ensuring the privacy and integrity of distributed systems. Interconnected devices have a huge potential of being exploited to gain access to broader networks. Designed digital spaces for exploration and collaboration introduce concerns about data privacy, virtual identity theft, and new forms of cybercrime. Cloud platforms offer scalability and convenience but can also pose several risks due to data breaches and misconfigurations. Addressing these challenges requires a proactive and adaptive approach to cybersecurity, leveraging innovative strategies to mitigate risks in the ever-evolving landscape of emerging technologies. Addressing these opportunities and challenges presented by emerging trends requires a collaborative effort from industry leaders, educational institutions, and policymakers. Organizations must invest in training and professional development programs to equip cybersecurity professionals, including women, with the necessary skills to leverage emerging technologies effectively. Fostering inclusive environments and providing equal opportunities for women promotes diversity and ensures that their perspectives are integrated into the design and implementation of cybersecurity strategies. By embracing the opportunities and addressing the challenges, the cybersecurity community can harness the full potential of emerging trends to strengthen digital security, mitigate threats, and create a safer digital environment for individuals, organizations, and society.

5.3 - Nurturing the Next Generation

The significance of nurturing the next generation of cybersecurity professionals is essential. As the digital landscape expands and cyber threats become increasingly sophisticated, there is a growing need for a diverse and skilled workforce to safeguard individuals, organizations, and societies from malicious activities. Numerous initiatives have emerged to introduce young learners to the field, which emphasizes the importance of cybersecurity awareness and education. These initiatives often include age-appropriate educational resources, workshops, and interactive programs designed to cultivate interest and promote cybersecurity knowledge and skills among students. Competitions are vital in engaging and motivating young learners to pursue cybersecurity careers. Capture-the-flag (C.T.F.)

competitions, hackathons, and cyber defense challenges provide hands-on experiences that test problem-solving abilities, technical skills, and teamwork. These events foster a spirit of healthy competition while nurturing talent and identifying promising individuals for future cybersecurity roles. Science, Technology, Engineering, and Mathematics (S.T.E.M.) education initiatives have embraced the inclusion of Cybersecurity as an essential component. By integrating cybersecurity concepts into existing S.T.E.M. curricula, educators can inspire students to explore the field, highlighting the interdisciplinary nature of Cybersecurity and its connections to science and technology.

Mentorship programs pair young learners with experienced cybersecurity professionals who serve as role models and guides. These programs provide valuable insights, career guidance, and support to aspiring cybersecurity professionals, helping them navigate educational pathways, gain practical experience, and develop the necessary skills to succeed. Collaboration between educational institutions, industry partners, and cybersecurity organizations is crucial in nurturing the next generation of professionals. Through internships, apprenticeships, and partnerships, students gain real-world exposure, industry insights, and hands-on experience, bridging the gap between academia and industry requirements. This sacred relationship should be based on collaboration, trust, and mutual respect within the cybersecurity community. It exemplifies the spirit of support and camaraderie that should exist within the industry, fostering an environment where individuals can thrive and excel. In the ever-evolving landscape of Cybersecurity, mentorship remains a cornerstone of success. Embrace the power of mentorship as a mentor and a mentee to unlock the potential to make a lasting impact on the field. Together, we can shape the future of Cybersecurity and create a world where individuals and organizations are empowered to navigate the digital realm with confidence and security.

By nurturing the next generation of cybersecurity professionals, we empower young learners to become the defenders of the digital realm. Initiatives, programs, and organizations dedicated to promoting cybersecurity education and engagement among young individuals are instrumental in fostering interest, cultivating talent, and developing the skills needed to protect our digital crowns.

Through these efforts, we can build a diverse and resilient cybersecurity workforce capable of addressing the ever-evolving cyber threats of tomorrow.

The Reign of Women in Cybersecurity

The core themes explored throughout the book start with the world of cybersecurity, the importance of women's participation in the field, the challenges they face and continue to overcome, their inspiring success stories, strategies for advancement and empowerment, and the significance of diversity and inclusion. The book is intended to educate and inspire all to pursue careers in cybersecurity, especially women, as numerous opportunities and exciting prospects are awaiting those who embark on this path. From protecting critical infrastructure to defending against emerging threats, women are encouraged to recognize the immense impact they can make in securing the digital realm.

Women's voices, perspectives, and contributions are extremely important in shaping the future of cybersecurity. Diversity and inclusion strengthen the field by driving innovation and creating more resilient defenses against cyber threats. Encouraging women to bring their unique talents and insights to the table inspires them to envision themselves as catalysts for positive change and transformation. By embracing their crowns, women are empowered to identify their aspirations and passions to break barriers, challenge gender biases, and seize the opportunities that await them in the cybersecurity domain.

6.1 - Encouraging More Women to Join the Field

Cybersecurity is rapidly evolving, driven by emerging technologies and the ever-evolving cyber landscape. In this context, the critical role of women in shaping the future of the industry cannot be overstated. Women bring unique and diverse perspectives to the field by enriching the collective intelligence of the industry. Their unique experiences and backgrounds foster a broader understanding of cyber threats, vulnerabilities, and mitigation strategies. By embracing diversity, the field can develop more robust and comprehensive approaches to cybersecurity that address various challenges.

By embracing diverse ideas and encouraging women's participation, the industry can tap into broader insights and approaches. Creativity is a valuable asset in cybersecurity, as it allows for exploring new methodologies, technologies, and strategies to address evolving threats. Women's fresh perspectives can help uncover novel solutions and enhance existing practices. Moreover, collaborative problem-solving is crucial in tackling cybersecurity challenges effectively. Women often excel in teamwork and communication, which fosters an environment of collaboration, knowledge sharing, and support. These skills contribute to more comprehensive and resilient cybersecurity strategies and stronger defense mechanisms. Promoting gender diversity and inclusion in a male-dominated industry creates opportunities for women. It benefits the industry by leveraging a diverse talent pool and fostering an environment encouraging innovation, collaboration, and success.

The emergence of new technologies such as artificial intelligence, machine learning, and blockchain brings opportunities and challenges to cybersecurity. Women's active participation and leadership in these areas are vital for ensuring that the future of the industry is shaped with inclusivity and a deep understanding of the potential risks and benefits of these technologies. By actively encouraging more women to pursue careers in cybersecurity, we can inspire and empower the next generation of professionals. Visible role models and mentors are crucial in nurturing young talent and challenging societal stereotypes. By creating pathways and opportunities for women to excel, we create a ripple effect leading to increased diversity, inclusivity, and innovation within cybersecurity.

To harness the full potential of women in shaping the future of cybersecurity, organizations, and the industry must take proactive steps. This includes implementing diversity and inclusion initiatives, promoting equal opportunities for career advancement, and creating supportive networks and mentorship programs. Encouraging young girls to explore S.T.E.M. fields and providing educational resources can also help bridge the gender gap and inspire future generations of women to pursue cybersecurity careers. By recognizing and embracing the critical role of women in the industry, we can create a more secure, resilient, and inclusive future. Their unique skills, perspectives, and contributions will play a pivotal role in addressing the challenges posed by emerging technologies, fostering innovation, and safeguarding our digital landscape for generations to come.

6.2 - Embracing the Power of U.N.I.T.Y.

Take a moment to dream of what the future of a career in cybersecurity could look like. In a realm where the boundaries of the digital world merge with our realities, a new narrative unfolds—one that celebrates the power of unity between women and men. Women converge to create a powerful force of unity where Wisdom, Opportunity, Mentorship, Empowerment, and Nurturing reign supreme. The journey towards breaking barriers, embracing diversity, and empowering women to thrive in a cybersecurity career is now a reality. Cybersecurity professionals gain wisdom from their experiences for any opportunities (i.e., emerging technologies or global events). The guidance mentors provide will empower us to reach new heights, all while nurturing the support system that fosters growth and success.

Individuals from all walks of life, with diverse backgrounds and unique strengths, come together to protect the cybersecurity realm. They embrace the principles of Unison, recognizing that their collective efforts are stronger than any individual's. Through networking, they forge connections and build bridges, forming a web of collaboration across organizations, industries, and borders. Inclusivity is their guiding light, ensuring every voice is heard, valued, and included. They understand that true progress can only be achieved through teamwork, combining their expertise, ideas, and skills to overcome challenges and find innovative solutions. And in the spirit of Yearning, they embrace the deep desire for progress, yearning to bridge gaps and break barriers, recognizing that the unity of men and women working together in harmony is the key to maintaining the security of our digital world.

Welcome to the Cybersecurity Queen's dream, where the strengths of all are harnessed for a safer and more inclusive future. Can you envision a future where women and men stand side by side, wielding their collective strengths to protect the digital crown? A future where diversity, collaboration, and inclusivity reign, paving the way for innovative solutions, enhanced security, and a harmonious balance in the cyber realm. Can you envision the future where women shine brightly, empowering the cybersecurity landscape with their knowledge, skills, and unique perspectives? The future is in our hands, and together, we can shape a world where the collective power of all safeguards the digital crown.

Everyone should feel encouraged to forge their paths, embrace the challenges, and make their mark in the cyber kingdom. Reflect on your passions, aspirations, and potential to determine how they align with opportunities in the cybersecurity realm. Individuals are inspired to step forward, embrace their crowns, and shape the future of cybersecurity with confidence and conviction. The power to make a difference lies within each person who dares to dream and pursue a career in cybersecurity. Let this book be a guide, a source of inspiration, and a reminder of the strength and potential within each of us. The path to a career in cybersecurity may present challenges, but know that you have the resilience, determination, and talent to overcome them. Embrace the opportunities to continually learn, adapt, and grow as the field evolves. Embrace the power of diversity and inclusion, knowing that your unique perspectives and experiences will strengthen the industry and drive innovation.

The time is now to take your place in the cyber kingdom. Be bold, courageous, and confident in your abilities. You have the power to shape the future of cybersecurity and make a lasting impact. Embrace your crown and let it symbolize your strength, resilience, and unwavering commitment to creating a safer digital world for all. Each voice and contribution are valuable and essential in creating a safer digital future. By embracing our crowns, women and men can step into their power and make a profound impact in the cybersecurity domain. Remember, we're in this together! Find supportive communities, mentorship programs, and organizations dedicated to empowering women in cybersecurity. Seek out these resources, build strong networks, and surround yourself with allies who believe in your potential. Let us build a more diverse, inclusive, and secure digital world together.

www.ingramcontent.com/pod-product-compliance
Lightning Source LLC
Chambersburg PA
CBHW071743090426
42738CB00011B/2552